MIND BLOWING! THE BRAIN

EMOTIONS

by Joyce Markovics

Cherry Lake Press
Ann Arbor, Michigan

Cherry Lake Press

Published in the United States of America by Cherry Lake Publishing Group
Ann Arbor, Michigan
www.cherrylakepublishing.com

Reading Adviser: Beth Walker Gambro, MS Ed., Reading Consultant, Yorkville, IL
Content Adviser: Heather A. Lau, MD MS, Pediatric Neurologist
Book Designer: Ed Morgan

Photo Credits: © freepik.com, cover and title page; © freepik.com, TOC; © Master1305/Shutterstock, 4; Wikimedia Commons, 5 top; Wikimedia Commons, 6; © freepik.com, 7; © freepik.com, 8; © freepik.com, 9; © freepik.com, 10; © freepik.com, 11 right; © freepik.com, 11 left; Wikimedia Commons, 12 top; © SciePro/Shutterstock, 12 bottom; © freepik.com, 13; © Dmitry Burlakov/Shutterstock, 14-15; © freepik.com, 15 right; © Anastasia Tveretinova/Shutterstock, 16; © Roman Zaiets/Shutterstock, 17; © freepik.com, 18 left; Geoff B Hall, Wikimedia Commons, 18 right; © freepik.com, 19; © michaeljung/Shutterstock, 19 bottom; Jack and Beverly Wilgus, Wikimedia Commons, 20; Wikimedia Commons, 21 left; Wikimedia Commons, 21 right; © freepik.com, 22.

Cherry Lake Press is an imprint of Cherry Lake Publishing Group.

Library of Congress Cataloging-in-Publication Data
Names: Markovics, Joyce L., author.
Title: Emotions / by Joyce Markovics.
Description: Ann Arbor, Michigan : Cherry Lake Publishing, [2023] | Series: Mind blowing! the brain | Includes bibliographical references and index. | Audience: Grades 4-6
Identifiers: LCCN 2021035272 (print) | LCCN 2021035273 (ebook) | ISBN 9781534199538 (hardcover) | ISBN 9781668900673 (paperback) | ISBN 9781668906439 (ebook) | ISBN 9781668902110 (pdf)
Subjects: LCSH: Emotions–Juvenile literature. | Emotions–Physiological aspects–Juvenile literature.
Classification: LCC QP401 .M33 2023 (print) | LCC QP401 (ebook) | DDC 612.8/232–dc23
LC record available at https://lccn.loc.gov/2021035272
LC ebook record available at https://lccn.loc.gov/2021035273

Printed in the United States of America
Corporate Graphics

CONTENTS

EMOTIONS 101

As you read the words on this page, what are you feeling? Are you bored, excited, or confused? Maybe you're experiencing an entirely different emotion. So what exactly are emotions—and where do they come from?

Hundreds of years ago, people thought emotions sprang from different **organs** in our bodies. They believed the liver caused anger, while the spleen brought about sadness. The heart—you guessed it—was a source of love.

The word *emotion* can be traced to 1579. It comes from the French term émouvoir, which means "to stir up."

In the 1600s, doctors looked inside the body. They saw blood pumping through the heart. Over time, they began to understand how the different organs worked. Eventually, they learned that emotions come from one incredible organ—the brain!

William Harvey was the first doctor to describe how the heart pumps blood to the rest of the body around 1618.

Emotions are created by your brain. Scientists describe anywhere from 6 to 27 **distinct** human emotions. Some emotions can feel good, such as joy or love. Others can feel bad, like anger or sadness. Yet there is no such thing as a good or bad emotion.

Emotions helped early humans survive threats. For example, they prepared people to either face or run from danger.

Can you tell how people are feeling just by looking at their faces? People used to think so. They even drew pictures showing how faces express various emotions. But according to brain scientist Lisa Feldman Barrett, no one can really tell how another person is feeling.

You are the only person who knows how you feel. Lisa describes emotions as "guesses your brain constructs." Emotions give us information and help us decide how to react to the world. By learning about them and the brain, we can better understand ourselves and others!

Every person on the planet has a unique brain and emotions.

INSIDE YOUR BRAIN

Your brain is a squishy pink organ inside your skull. It's made up of billions of cells working together. Some of these cells, called neurons (NOO-ronz), have an important job. They carry signals all over your brain and body, allowing you to move, think, and feel.

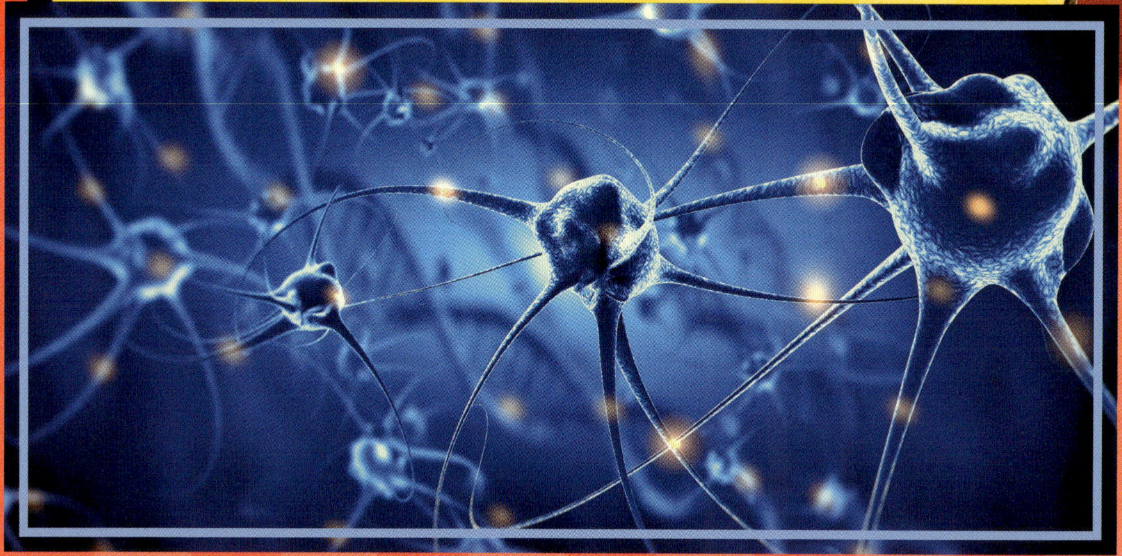

Neurons have long, spidery arms. A bundle of neurons is called a nerve.

Every second, millions of signals zoom through your brain. They pass along information to the rest of your body. Maybe you have an itch or see something scary. Signals from your brain make it possible for you to scratch that itch or experience fear. As this happens, different areas of your brain are **activated**.

An adult human brain weighs about as much as a cantaloupe! And it has the same texture as creamy custard.

Your brain has about 100 billion neurons. Neurons connect with each other using electrical and chemical signals. These signals can travel up to 250 miles (402 kilometers) per hour!

The brain has many areas that each control a different part of our body or personality. The section most closely linked to our emotions is the **limbic system**. It's buried deep within the brain, under the cerebral cortex (suh-REE-bruhl KAWR-teks) and above the **brain stem**. The limbic system includes the amygdala (uh-MIG-duh-luh), **hypothalamus**, **hippocampus**, and a few other key parts.

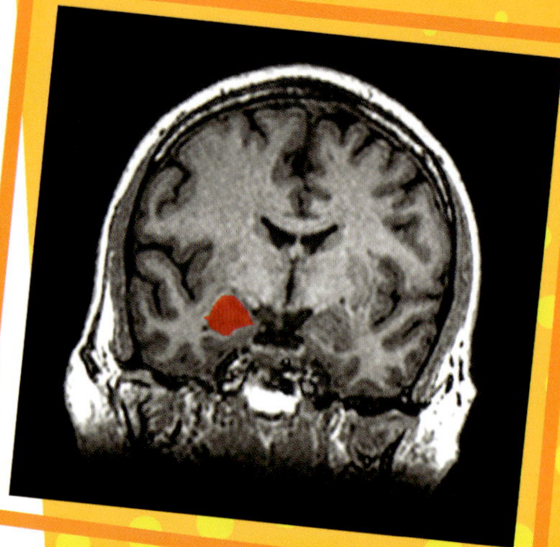

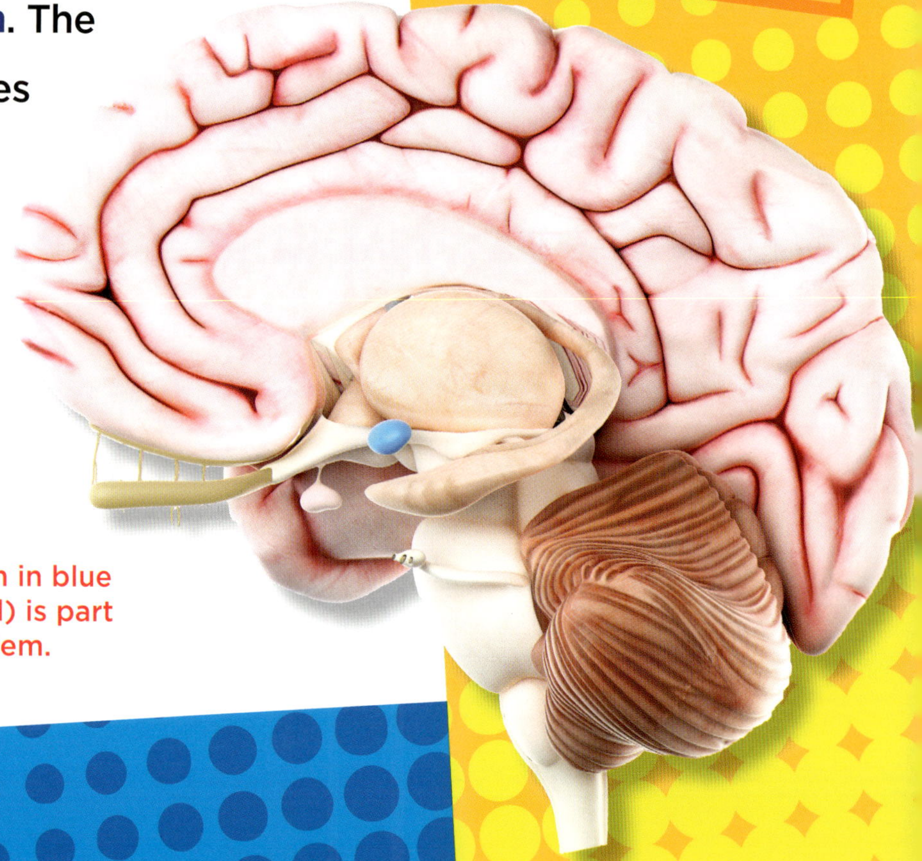

The amygdala (shown in blue here and above in red) is part of the limbic system.

The amygdala is a pair of almond-shaped parts in the brain. It plays an important role in how we deal with pleasure, fear, and anger. It helps us figure out who is a friend or enemy and how to respond accordingly.

The cerebral cortex is about as thick as cardboard. It's the wrinkly, pinkish-gray outer layer of our brains. It connects to the limbic system and is involved in everything we do.

If you're in danger, the amygdala sparks the hypothalamus. Then the hypothalamus sends signals to **glands** that make **hormones**. Hormones enter your blood, causing sweaty armpits, a pounding heart, or other changes. These responses alert you to danger and help you stay safe.

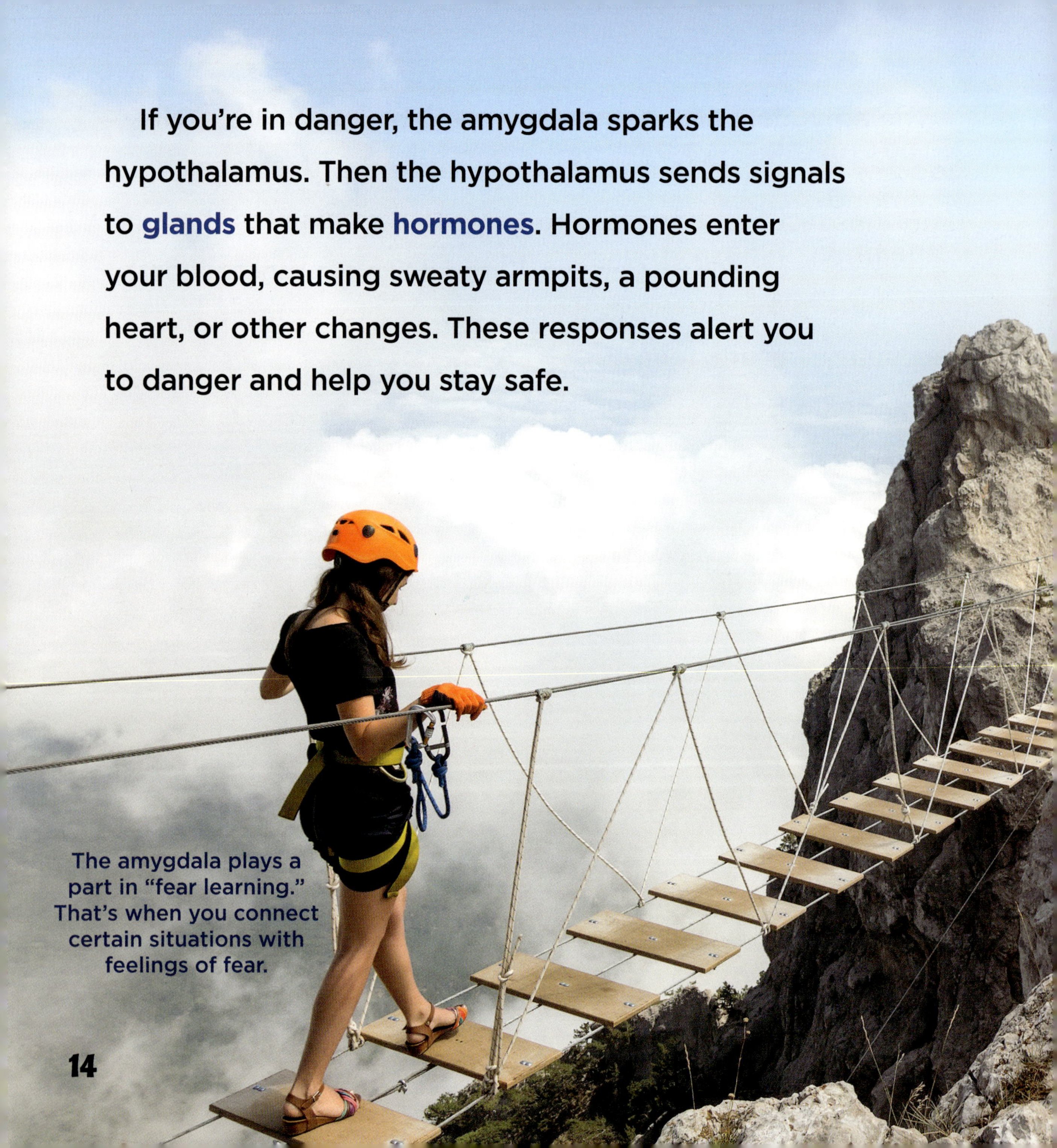

The amygdala plays a part in "fear learning." That's when you connect certain situations with feelings of fear.

The amygdala also helps us remember—especially really good or really bad things. Memories linked with strong emotions last the longest. The amygdala works with the hippocampus. Shaped like a curvy seahorse, the hippocampus is the memory center of your brain.

The body's reaction to fear is called fight or flight. People either face or flee danger.

In order for signals to travel through the brain and body, neurons release neurotransmitters. These chemicals transmit messages between neurons and muscles.

BRAIN EXPERIMENTS

To learn more about emotions and the brain, scientists perform tests. They use scanners such as MRIs as well as other tools to look inside the brain. Sometimes they safely **stimulate** the brain with electricity. In one experiment, scientists sent weak electric charges into the hypothalamus of a live cat.

Stimulating one tiny area of the brain caused the cat to go from friendly to angry. Sending charges into other parts of the cat's brain caused different reactions. Similar experiments have been done on monkeys and humans.

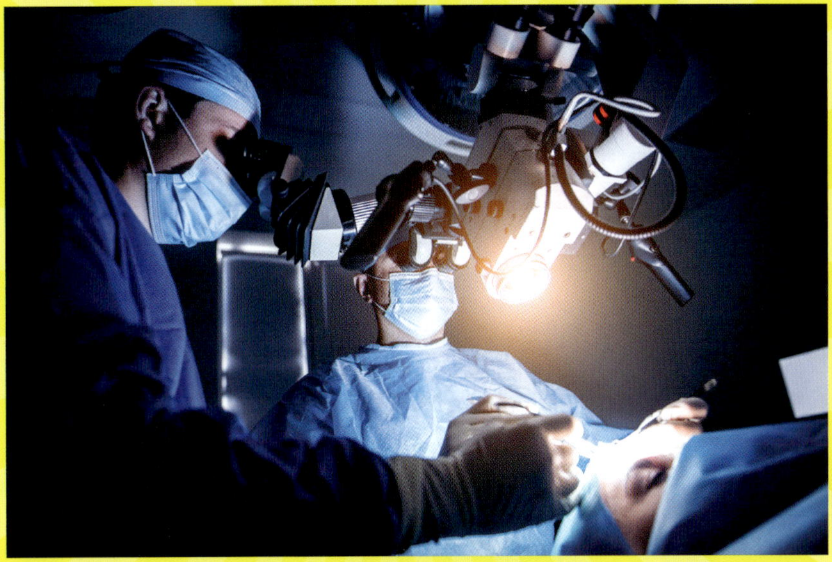

The brain doesn't feel pain, so people are sometimes kept awake during brain surgery. This helps doctors make sure they're not harming important areas of the brain.

Despite many experiments, scientists are still not sure how some parts of the brain work. For example, we still don't know how some kinds of neurons work.

In 2015, scientists discovered that a brain part called the precuneus (pree-KYOO-nee-uhs) may have a lot to do with happiness. People with more tissue, or **gray matter**, in this area were happier. Experts think that the precuneus helps transform information into happy feelings.

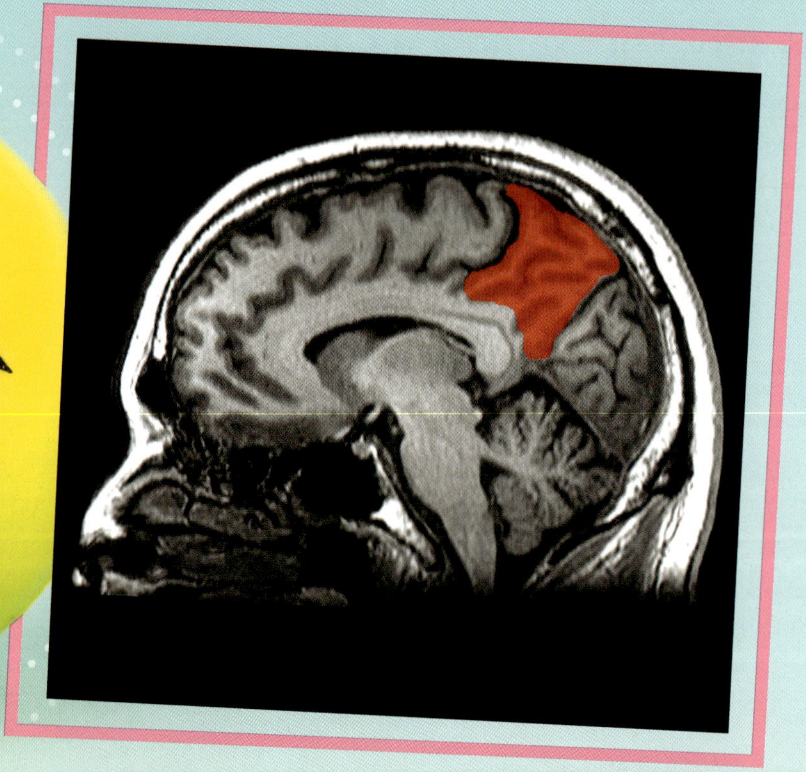

The red area is where the precuneus is located.

What's fascinating is the brain "remembers" feeling happy. When you think about an experience that made you happy, you might be flooded by a feeling of happiness. The brain also connects other past experiences to feelings, such as anxiety. For example, a past experience could make you feel nervous whenever you take a test.

What is love? This intense feeling comes from hormones in the brain. The main one is oxytocin (ok-si-TOH-suhn). It increases when you hug or kiss someone!

Dopamine (DOH-puh-meen) is a neurotransmitter. Your brain releases it any time you're expecting to feel good!

A CURIOUS CASE

Sometimes our brains have a disease or are injured. In 1848, a railroad worker named Phineas Gage was in an awful accident. After an explosion, he was struck by a long iron bar that passed through his skull and brain. Amazingly, Phineas lived. But he was "no longer Gage," according to his friends.

Phineas Gage holding the bar that speared his skull

Phineas went from being quiet and gentle to outspoken and angry. Later, scientists found out that the part of his brain that was damaged is critical for processing emotions. As it turns out, studying the brain when it isn't working properly helps us learn even more about it!

Phineas's damaged skull

BRAIN GAMES

Did you know that your brain signals can match another person's? This is called coupling or **synchronization**. We pick up on a person's body language and facial expressions. Then our own brains mirror that person's emotional response!

- Look at the photo or watch a video of someone crying. Focus on the person's face.
- Or think back to a time when you were with someone who was really upset. How did you feel? Were you a little sad yourself? Why do you think that is?

GLOSSARY

activated (AK-tuh-VAY-tuhd) made something function

brain stem (BRAYN STEHM) the part of the brain that controls essential actions, like breathing and swallowing

chemical (KEH-muh-kuhl) a natural substance that helps the body function

distinct (dih-STINGKT) not the same

electrical (i-LEK-truh-kuhl) related to the flow of electricity, a form of energy

glands (GLANDZ) body parts that produce natural chemicals

gray matter (GREY MAT-uhr) the darker tissue of the brain and spinal cord that's made up of nerve cells

hippocampus (hih-puh-KAM-puhs) the part of the brain that's mainly involved in memory

hormones (HOR-mohnz) chemicals that are made by certain glands

hypothalamus (hye-poh-THAH-luh-muhs) the part of the brain that's mainly involved in emotions

limbic system (LIM-bik SIH-stuhm) the region of the brain that deals with emotions and memory

organs (OR-guhnz) body parts that do particular jobs

stimulate (STIM-yuh-layt) to raise the levels of something in the body

synchronization (sing-kruh-nuh-ZAY-shuhn) when two or more things work together at the same time

FIND OUT MORE

Books

Mason, Paul. *Your Mind-Bending Brain and Networking Nervous System*. New York, NY: Crabtree Publishing, 2016.

Silver, Donald M., and Patricia J. Wynne. *My First Book About the Brain*. Mineola, NY: Dover Publications, 2013.

Simon, Seymour. *Brain: Our Nervous System*. New York, NY: HarperCollins, 2006.

Websites

American Museum of Natural History: Brain Introduction
https://www.amnh.org/exhibitions/brain-the-inside-story/brain-introduction

The Franklin Institute: Your Brain
https://www.fi.edu/your-brain/interactives

KidsHealth: Understanding Your Emotions
https://kidshealth.org/en/teens/understand-emotions.html

INDEX

ABOUT THE AUTHOR

Joyce Markovics has written hundreds of books for kids. She's fascinated by the human brain and all its complexities. She would like to dedicate this book to her sister Trisha, who has an extremely compassionate brain.